AF753192

Beaverbank

Between Material and Process

by *Frank Chiu*

The documentation of a project by the TUNS School of Architecture in collaboration with the Ceramics Department of the Nova Scotia College of Art and Design

Beaverbank
Between Material and Process
By: Frank Chiu
Editor: Richard Kroeker

TUNS Press
Faculty of Architecture
Technical University of Nova Scotia
P.O. Box 1000
Halifax, Nova Scotia
Canada B3J 2X4

 Published 1995
Printed in Canada

Cover painting, drawings and book design
by the author.
Printing: etc. Press Limited, Halifax

Canadian Cataloguing in Publication Data

Chiu, Frank, 1966-

Beaverbank

ISBN 0-929112-34-2

1. Architectural design. 2. Building.
3. Building materials. I. Title.

TH145.C54 1995 690 C95-950234-3

Contents

bark-skin

Grant MacEachern

Foreword

Richard Kroeker

Our culture has tended to create a separation between what we understand as the milieu of nature and the milieu of technology or artifice. Similarly, we struggle to maintain the empirical and the intuitive in their separate categories. Given our current understanding that natural forms and systems organize themselves according to a dynamic of optimization, there seems to be a strong basis for re-examining this historical definition of categories.

Architecture is the most primitive of the arts: primitive in the best sense. It is a response to the most primal of urges and therefore a useful vehicle for exploring these artificial groupings. Architecture can be understood as a material record of human technological achievement. In a context dedicated to the notion of progress, as measured by the distance covered from the point of origin, we risk losing contact with origins.

The Beaverbank project is one in a series of projects done with students at the TUNS School of Architecture over the past four years, which explore the notion that artifice might be ultimately understood as a manifestation of nature. Taken together, they all explore the potential for the literal identity of site and built form. Human agency is an aspect of nature which ultimately cannot transcend it. We perceive in nature those orders which suit our desired, temporary ends.

The *Algonquin canoe* was built according to a process dictated by material and use. Complex principles of structure and dynamic were revealed in its making. The *Eskasoni lodges*, which were built using materials from the site, had an even stronger literal identity with the site. The role of the designer was as the active agent, following instructions inherent in material, site and use. Our readings of these were helped by centuries of tradition and evolution.

Birchbark canoe built at the TUNS School of Architecture 1991, presently in the Maritime Museum, Halifax, Canada.

Eskasoni lodges (one shown) built with the instruction of John Henry Lafford and Barry Bernard of Eskasoni.

The Beaverbank project began with loosely defined parameters. A suitable area in the forest was identified. Within that area, the group chose a specific site and built a shelter using surrounding materials as building fabric. The intent was that the structure be permanent in the sense that it could be usable in all seasons and last for a period of years.

Much of what is built in Canada uses wood from the forest, and is built within the forest, or what was once the forest. Many of the processes by which we use the forest for building seem to be approaching obsolescence. The project was a response to this, questioning the ways in which our assumptions about technology have conditioned the current thinking on how wood is used in building.

The drawings done by Frank Chiu of the structure are an integral part of the process. The documentation is an abstraction of the work after the fact which reveals new ideas and opens new lines of inquiry.

Neil Forrest

As part of an ongoing educational experiment with students and faculty at the TUNS School of Architecture and NSCAD Ceramics, the ceramics department was invited to collaborate on the Beaverbank project.

The complexities of our respective disciplines required a new dialogue — finding common ground in our specialized languages and approaches. Each group, ceramists and architects, had to understand the other with their unfamiliar thought processes and work environments. The ceramists contributed architectural elements that increased the usability and artistry of the structure. Light "cannons", rain catchers and a central hearth and chimney create focal points that stimulate the dialogue of interior/exterior.

The Beaverbank project reminds us of the distant traditions of craft and architecture in which the role of maker and designer is merged. The emphasis on learning at the site and engagement in direct building methods, brought instant questions and fresh insights. The process became an oracle of experience. We asked ourselves how is it possible to wrestle issues of design, performance and artistry while holding the very clay and timber that is Beaverbank?

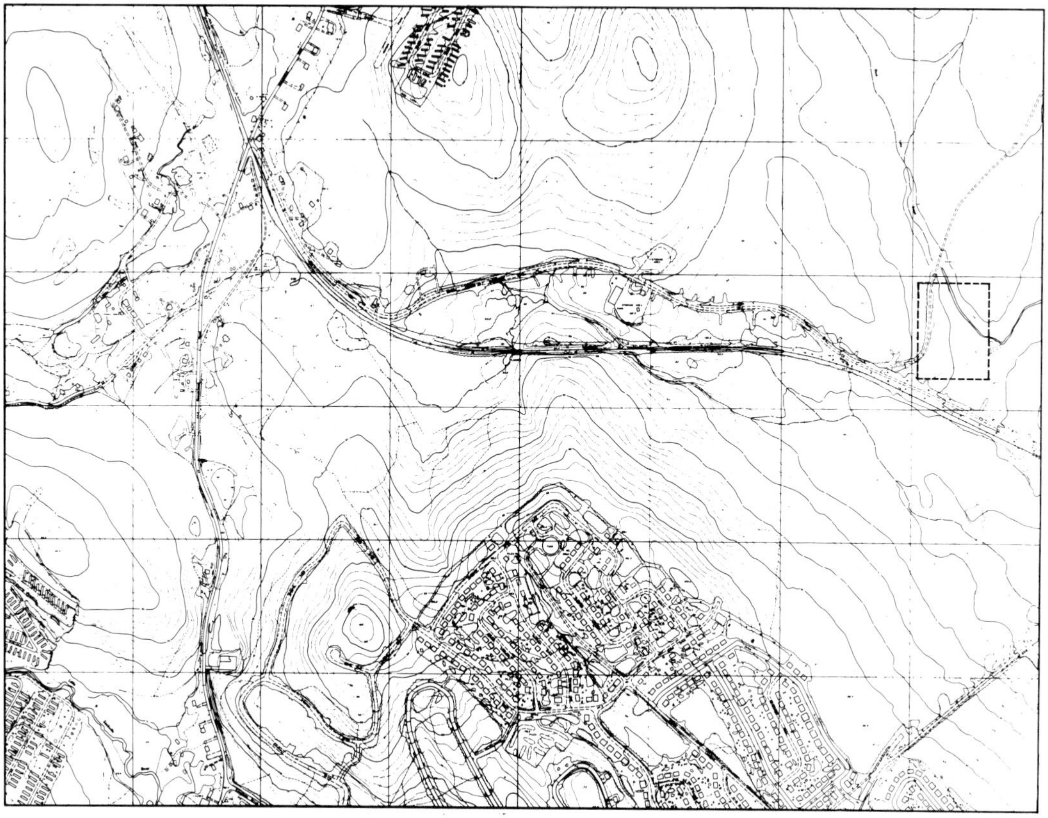

Context - North of Middle Sackville, the site exists within the forest.

Beaverbank

Between Material and Process

When confronted by the presence of the woods, investigations require a respectful demeanour. The uninformed cutting of a tree can be interpreted as a negligent act against nature in which a living resource is violated. The Beaverbank project, however, pursued an appropriate and timely approach to material use in which the intrinsic value of the forest was given proper expression.

Over a three month period, a relatively permanent structure was developed by graduate students from TUNS, in collaboration with ceramists from NSCAD who created the central hearth and lighting elements for the building. The project was situated near an earlier construction, the *Bartel bridge*, which was built over the local stream. The basic intention was to create a retreat which offered protection from the elements, while at the same time enhancing our appreciation of the elements.

In beginning, the process drew on the way artisans gain inspiration from working directly with their medium. Just as the ceramic elements were developed directly in clay, the construction project went quickly to site after initial design ideas were tested with a few study models and full-scale explorations. The building design was developed as a group, working directly with the material and not through drawings. Using this approach, delight in the material was gained through hands-on play rather than speculation, which shaped the way of working.

Fundamental principles developed by the group established that material from the forest could be utilized through non-industrial processes using simple hand tools and ingenuity. *Red Spruce* was the most abundant indigenous species of tree at the site, and saplings from three to six inches in diameter were chosen for developing the bent-wood structure. Fittingly, trees of this size are not commercially viable, and through selective cutting the woods can grow stronger. By utilizing this neglected forest product, a readily-available source of renewable material was maximized.

The construction strategy cast aside prescribed conventions as improvisation became the rule. While experimenting, forceful attempts to bend a sapling caused it to break. Realizing the limitations of the material, actions were adjusted accordingly to allow the material to express its potential. Trying to lash the frame together with twine proved unsuitable so a decision was made to use stainless steel strapping when connections had to hold. Throughout, the procedure for making did not deny the existence of contemporary building developments but rather, the details evolved to find appropriate ways to include them.

In addition to the *Bartel* bent-wood bridge research, the methodology of the Beaverbank project was influenced by other projects developed at TUNS. Works such as the *Algonquin canoe* construction and the *Eskasoni lodges* provided insight into the development of an attitude of learning through direct material exploration. This body of reference, with its debt to the learned material traditions of Native American building, provided the Beaverbank work with its particular point of departure.

Locating

Site - Earthly elements of stone, wood, water and wind.
The bridge from a previous time providing reference for movement between building and circumstance.

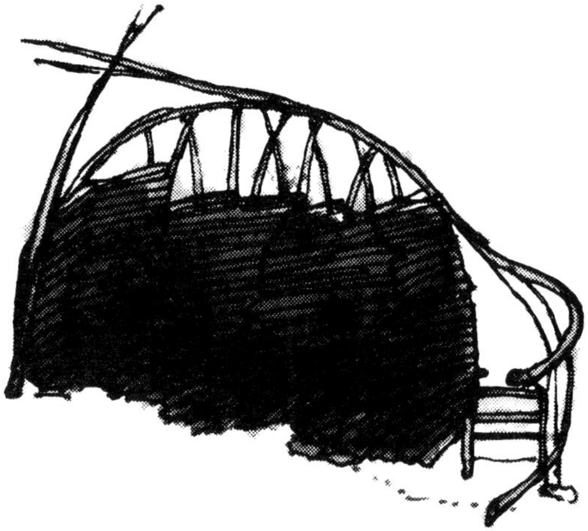

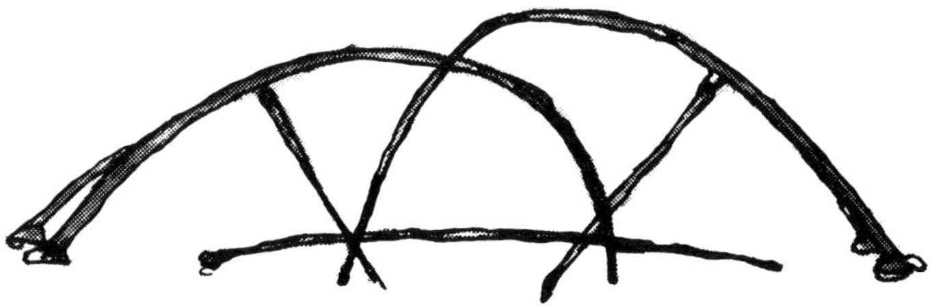

Structural Armature

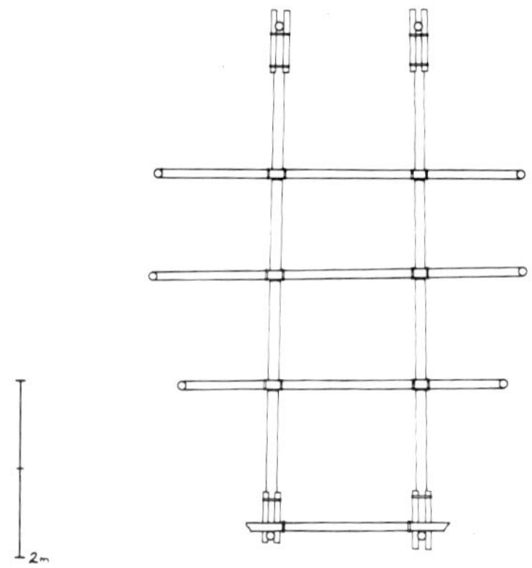

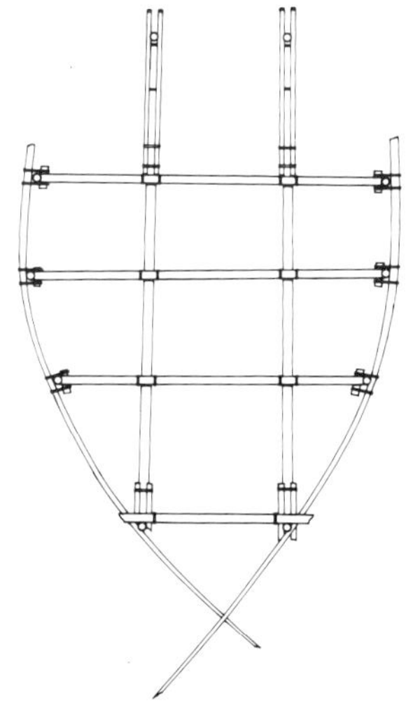

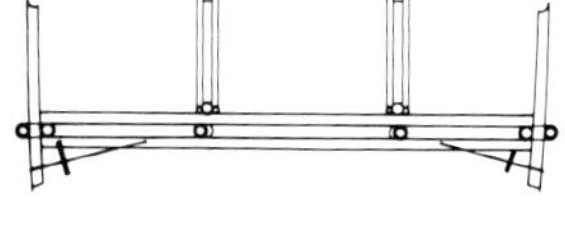

Platform - A horizontal grid-work made from rigid trees for resistance to anticipate anchoring.

- Plan and section drawings reveal the process of construction.

late January

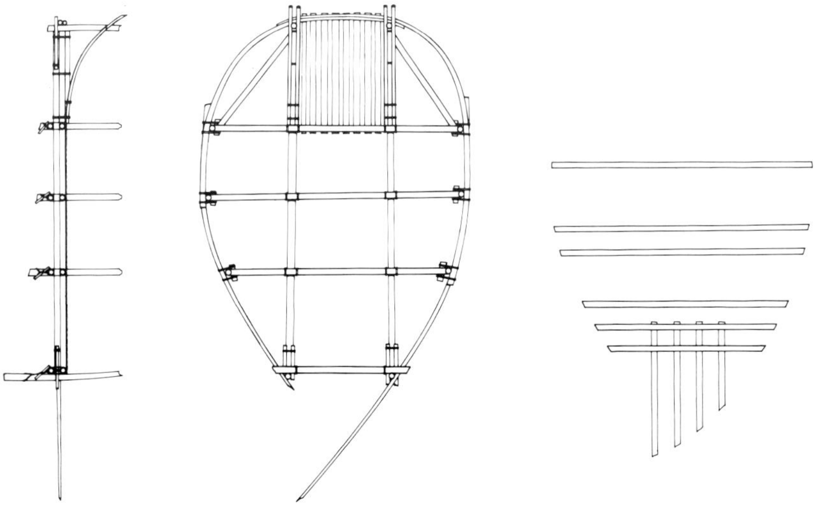

Early Bending - An arc defined by the material, expressing its potential.

15

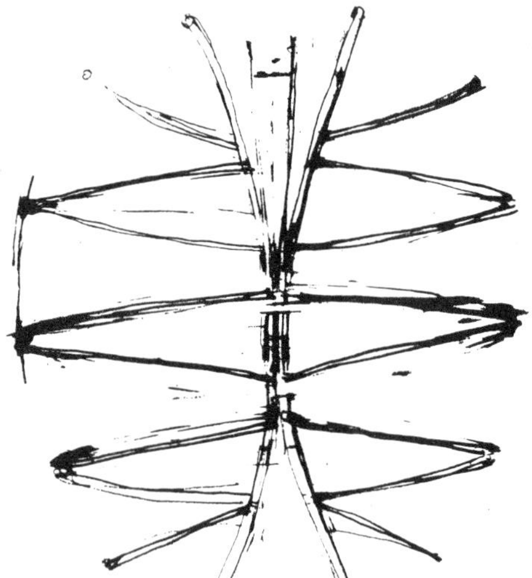

Spine - Verticals bending towards each other,
the central spine once formed
guides more to come towards it.

February

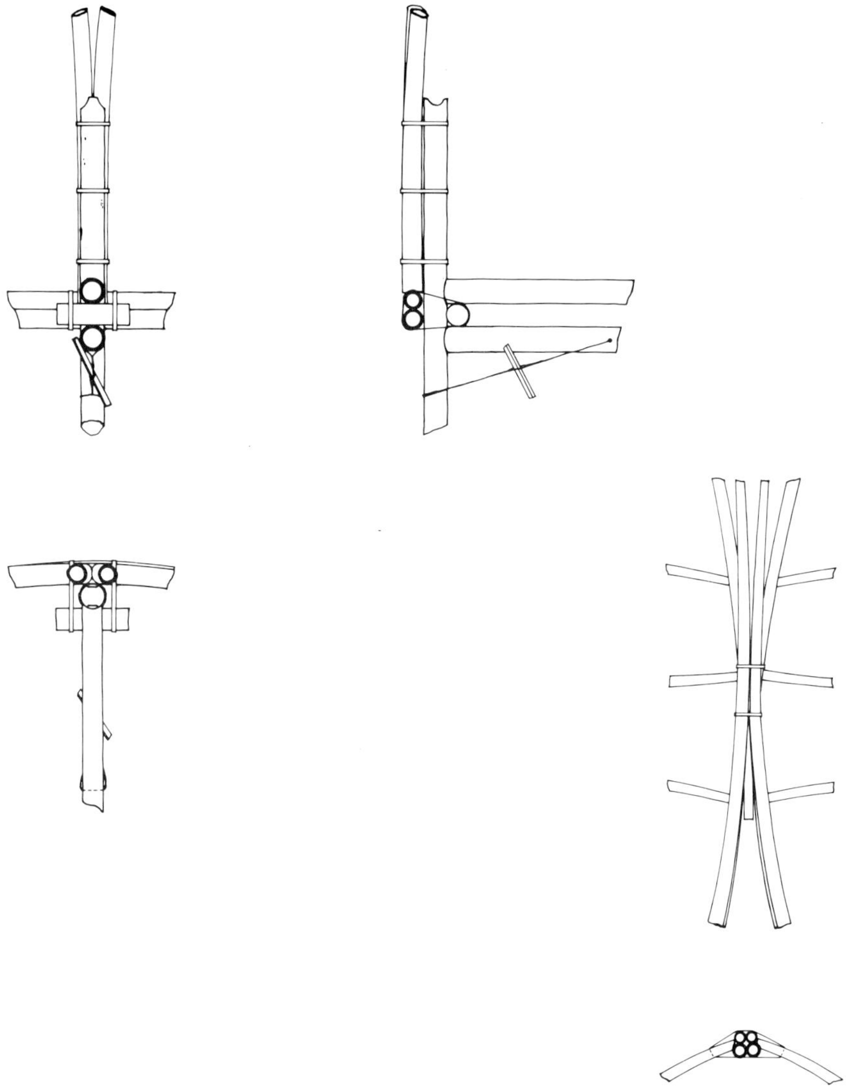

Tectonics - Adjacencies separated by perpendiculars, reinforced parallels vs. bending members, binding as connection.

- Details showing critical moments from anchoring point to spine.

February

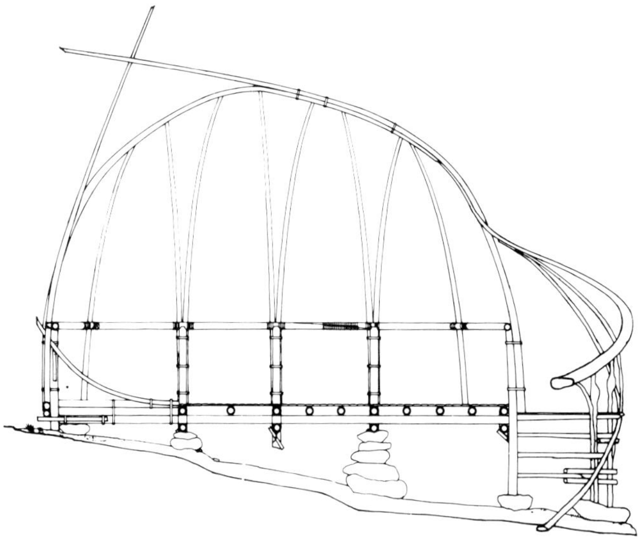

Structural Frame - Tectonics without prescribed orthodoxy
reveals a potent volumetric potential.

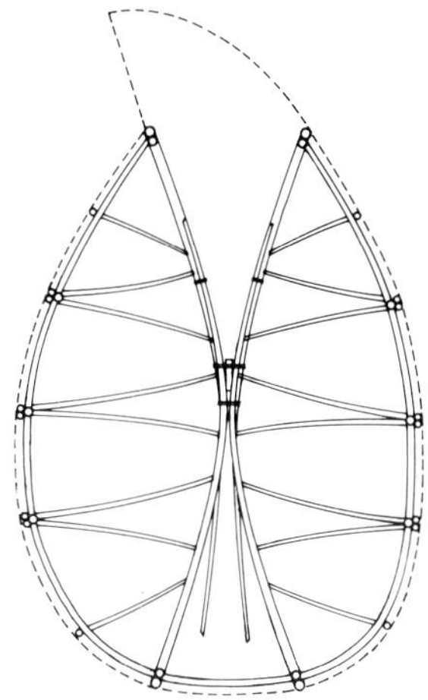

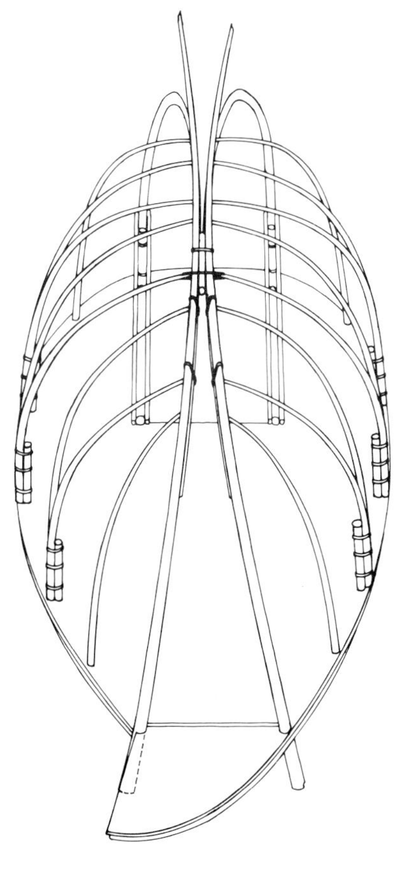

Skin(s)

Initial Wrapping - Covering/re-covering
a situation of extreme tension
mediated by reused bark.

March

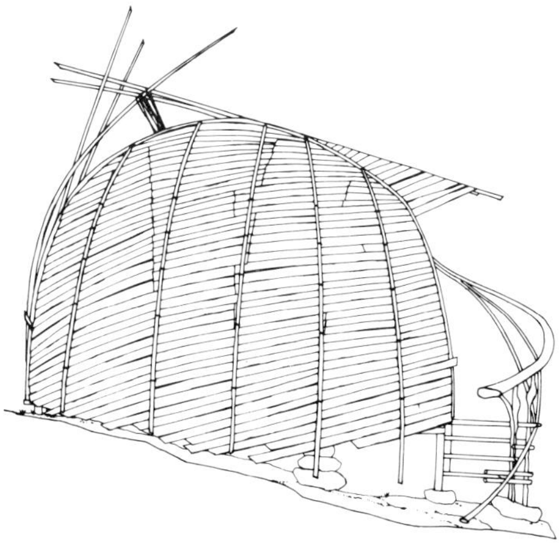

Outer Layering - A second skin of cedar shingling
embracing the elements
to weather and decay
revealing time.

25

March

Existing Stance

early April

Bearing

By taking a reading of existing positions in the field of architecture, the direction for movement in the Beaverbank project became clear.

In situating the work, interest gravitated towards traditional Native American building practices. Here, the cultural context of past achievements provided wisdom for survival in this age of diminishing resources. Native Peoples, such as the Mi'kmaq of the Cape Breton region, use their materials ingeniously to make essential articles for survival. Likewise, Beaverbank has endeavoured to develop self-reliant principles that do not subscribe to conventions of construction or form. The work addresses issues of material conservation, site ecology, and appropriate strategies for building that challenge preconditioned thinking in consumer culture.

As in the making of a bark canoe or a bentwood lodge, the Beaverbank work selected materials with care in order to conserve. The framing utilized sustainable, slender saplings that could be bent easily while the platform was constructed with thicker trees for stiffness. Bark, discarded from commercially processed timber, was salvaged for the initial skin of the structure which seemed poignantly appropriate. The outer shell, however, was layered with commercial cedar shingles that did not come from the site. The difficult decision to include this industrial product was the result of limitations in time and local resources. Once applied, the apparent toughness of the bark-skinned structure became more conventional and domesticated.

Acts of settlement involve site specific, environmental factors related to material and utility, sightlines, and protection from the elements. In locating the project, the existing *Bartel bridge* guided the group to a site above the sloped terrain by the stream's edge. Visual contact with the crossing was maintained from this sheltered position.

In the initial clearing of the land, the discovery of a high water table resulted in a decision to build the structure on footings. These supports were made from stacking large stones which were collected nearby. Intimate regard for the soil necessitated this non-invasive stance which respected the movement of run-off from the site towards the stream beneath. As the structure took form, the clearing of the site signified the flow between material and process, and mindful occupation of the land.

29

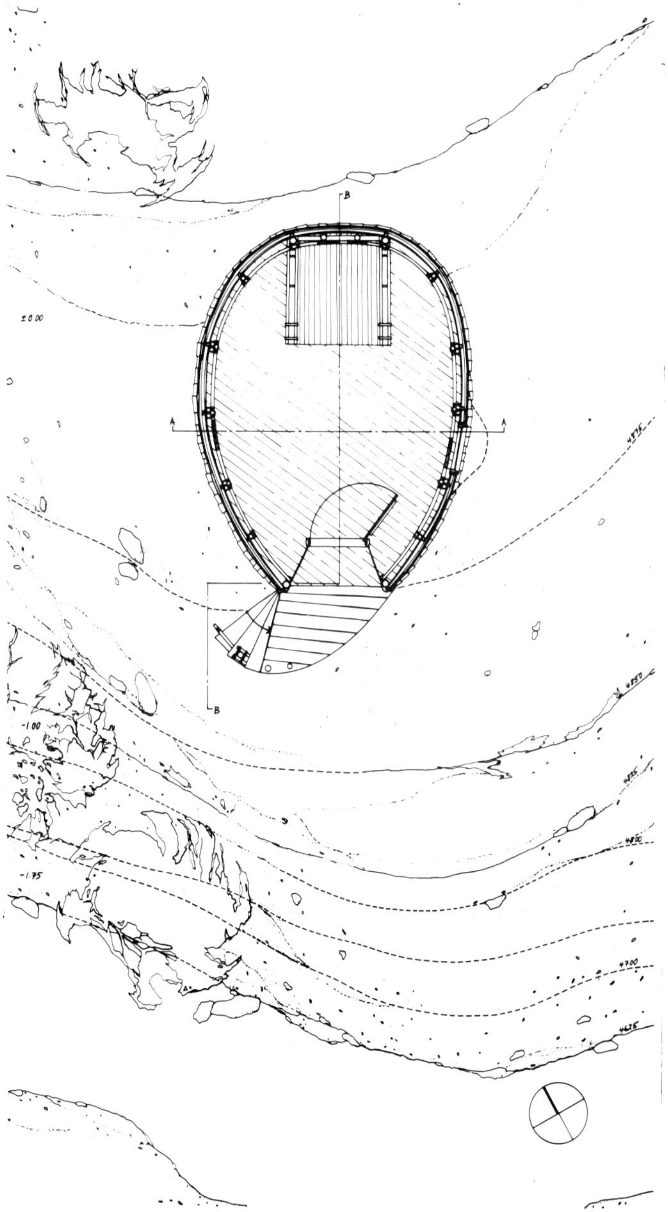

Final Documentation - Drawing to begin again.

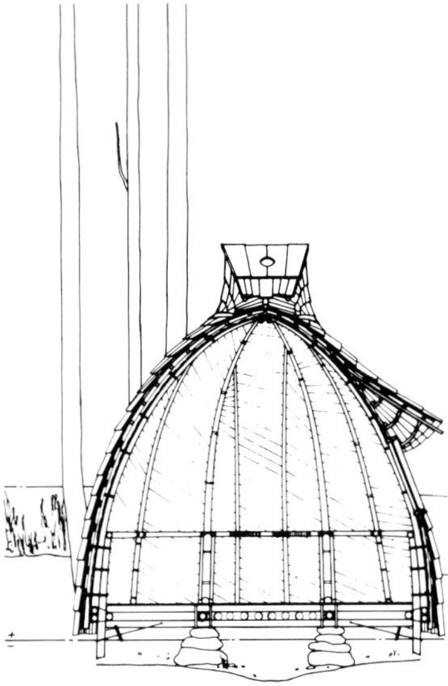

CROSS SECTION (AA)

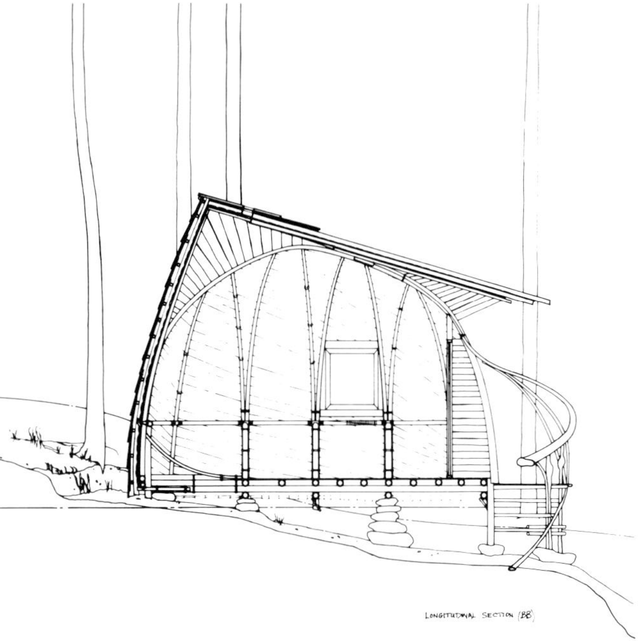

LONGITUDINAL SECTION (BB)

EAST ELEVATION

I like living close to nature
My ancestors did.
And being closer to the stars at night
And reading dreams
On interpretation, on what is right

I like living close to nature
My parents did
Meeting the sunrise at dawn
Upon seeing the sign of warmth
The sun song.

I like living close to nature:
I still do today.
Even improvising birch bark for a pot
To cook my meal.
The essence of my being original
In my instincts.

"On Being Original" by Rita Joe[1]

The aspiration of the Beaverbank project has been to liberate this spirit of originality when working towards appropriate material use between circumstance and building. The work endeavours to be sympathetic to past achievements and respectful of cultural traditions, yet ultimately contemporary. It is not about being nostalgic. Beaverbank demonstrates that knowledge is not to be separated from making; that through experience there is a lesson in the way we build which has direct bearing upon our views on architecture and the world at large.

1 Rita Joe is a Mi'kmaq elder and poet from Eskasoni. This poem is from her book, *Rita Joe: Lnu and Indians We're Called,* (Charlottetown: Ragweed Press, 1991), p.30.

Acknowledgements

Actions speak louder than words. The Beaverbank project involved a gathering of individuals whose resounding deeds for the undertaking are deserving of mention.

First and foremost I am grateful, as are my fellow classmates, to Professor Richard Kroeker. It was his vision to lead a group of us into the woods in the dead of winter to learn about architecture through building. Beaverbank is one in a series of material explorations which he has pursued with students. Richard's belief that architectural knowledge is not derived from theory alone but through the process of "making" will have a lasting effect upon the way we approach future tasks.

The project team, that I feel privileged to have been a part of and who are listed in closing, deserve full credit for all of their work. In particular, thanks must go to Francis LaPointe, our site coordinator, for his supportive attitude and boundless patience in building consensus within the group and for leading those of us with little experience in the art of hammer swinging to get the job done. It was Francis who was also instrumental in developing the initial idea for a building project in the woods and his foresight is deserving of much gratitude.

Additionally, a word of recognition is extended to Neil Forrest for revealing the wisdom of the ceramist's milieu, and to Barrett Lumber Company Limited for sharing their resources and for allowing us onto their property to experiment and discover new approaches to the use of traditional materials.

Regarding the publication, there are various contributors whose assistance has been invaluable.

Firstly, publishing is not a skill that one can learn overnight and the guidance provided by Donald Westin of TUNS Press has contributed immensely to this book. Another word of thanks goes to Professor Richard Kroeker for acting as Editor. His ability to focus on the essential aspect of a project has helped to keep the work on track. A special note of appreciation is also extended to Anne Margaret Copeland for her expert proof reading of the text and for making certain that my sentences made sense and did not run on, and on, and on, and on...

And finally, a school of architecture can only grow when it has teachers who not only show leadership, but also provide support. In this vane, I am grateful to Dean Frank Palermo for his constructive criticism and good advice. He shares our vision and was willing from the moment I proposed the idea of a publication to give his full backing, and that of the institution as well, towards making the work into something that can be shared.

F.C.

Credits

Project Coordinators

Richard Kroeker
Neil Forrest
Francis Lapointe

TUNS Group

Mark Archibald
Mark Atwood
Jeff Beanlands
Frank Chiu
Leif-Peter Fuchs
Leon Katsepontes
Nancy Labrecque
Margaret Hawkins
Doug Raphael
Martine Simard
John Stephens
Olexa Tkachenko
Brent Turchak
Pablo Yuste

NSCAD Group

Peter Eastwood
Christopher Garrison
Thomas Leung
Michael Lewis
Terrie MacDonald
Lisa Parsons
Lise Taylor
Heather Wilkenson

Photography

Richard Kroeker — p. 6.
Paul Toman — pp. 10, 26-27, 33.
Francis Lapointe — pp. 13-16, 19, 22(rt)-25.
The author — pp. 17-18, 22(lt), 32.

Photo-mechanical Transfers (PMTs) of the drawings were done by Bob Champion.

The *Bartel bridge* is part of a thesis project by Alvin Bartell.

The tree-bark print was created by Grant MacEachern.